Simple Science
Projects

PROJECTS WITH

AIR

By
John Williams

Illustrated by
Malcolm S. Walker

Gareth Stevens Children's Books
MILWAUKEE

For a free color catalog describing Gareth Stevens' list of high-quality children's books, call 1-800-341-3569 (USA) or 1-800-461-9120 (Canada).

Titles in the Simple Science Projects series:

Simple Science Projects with Air
Simple Science Projects with Color and Light
Simple Science Projects with Electricity
Simple Science Projects with Flight
Simple Science Projects with Machines
Simple Science Projects with Time
Simple Science Projects with Water
Simple Science Projects with Wheels

Library of Congress Cataloging-in-Publication Data

Williams, John.
    Projects with air / John Williams : illustrated by Malcolm S. Walker.
        p. cm. -- (Simple science projects)
    Rev. ed. of: Air. 1990.
    Includes bibliographical references and index.
    Summary: Provides instructions for making or using windmills, pinwheels, kites, and other devices that need air to set them in motion.
    ISBN 0-8368-0785-0
    1. Air--Experiments--Juvenile literature. 2. Wind power--Experiments--Juvenile literature. [1. Air--Experiments. 2. Wind power--Experiments. 3. Experiments.] I. Walker, Malcolm S., ill. II. Williams, John. Air. III. Title. IV. Series: Williams, John. Simple science projects.
    QC161.2.W56 1992
    620.1'07'076--dc20   91-50543

North American edition first published in 1992 by

Gareth Stevens Children's Books
1555 North RiverCenter Drive, Suite 201
Milwaukee, Wisconsin   53212, USA

Editor (U.K.): Anna Girling
Editors (U.S.): Eileen Foran
Editorial assistant (U.S): John D. Rateliff
Designer: Kudos Design Services
Cover design: Sharone Burris

Printed in Italy
Bound in the United States of America

1 2 3 4 5 6 7 8 9 97 96 95 94 93 92

CONTENTS

Words printed in **boldface** type appear in the glossary on pages 30-31.

AIR IS EVERYWHERE

Air is all around us. We cannot see it, taste it, or smell it, but we know it is there because we breathe it in and out all the time. When we breathe, our bodies use air to keep us alive.

These people are on a boat on a very windy day. They don't seem to mind the wind!

Make your own wind chart

When air moves in large amounts from place to place, it is called wind.

Go outside on a windy day. Run against the wind. Is it difficult? Now hold a cardboard shield in front of you as you run. Does the cardboard shield make running even more difficult?

Make a chart of the wind. Each day, look to see how windy it is and make a note of it on your chart. There might be just a light breeze, or there might be a **gale** or windstorm. Look for signs of the wind like this:

	Flags hanging straight down	Calm
	Laundry flapping	A breeze
	Trees swaying, branches broken off	A gale
	Trees blown down	A windstorm

WINDMILLS

We can use air in many ways. The wind can be used to drive windmills or to push boats along on the water. Modern windmills are used to make **electricity**.

This is a "wind farm," where electricity is made from the wind.

Make your own model windmill

You will need:

A cardboard box
Felt-tip pens
Cardboard
A paper fastener
Scissors

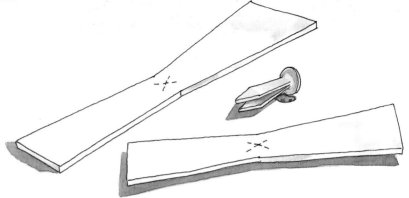

1. With felt-tip pens, draw a windmill on the box.

2. Cut out two strips of cardboard for the **sails**.

3. Use a paper fastener to attach the sails to your windmill.

Make your own pinwheel

You will need:

Construction paper
A pencil and ruler
Scissors

A strong pin
A plastic bead
A short wooden stick

1. Cut out a square of paper and draw **diagonal** lines across it, as shown.

2. Make a small hole in the middle, where the lines meet. Cut along the lines, about halfway to the middle.

3. Fold into the middle those corners marked with a cross.

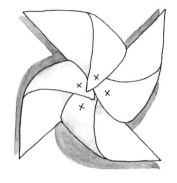

4. Stick the pin through each corner and then through the hole in the middle of the square. Push the pin through the bead and then attach it to the stick. Now try your pinwheel outside.

Further work

Try your pinwheel outside on different days. Make a chart to find the best wind strength for your pinwheel. On different days, count how many times the pinwheel turns around in five seconds. Color one of the sails red to help you count how many times the pinwheel turns.

DAY	WIND STRENGTH	NUMBER OF TURNS	COMMENTS
Day one	Calm	Zero to one	Moving slowly
Day two	Breeze	Five	Spinning fast
Day three	Strong Breeze	Eight	Spinning very fast

This pinwheel is like the one you can make, but with more sails. Make one like this, using two pieces of paper, one on top of the other.

WIND WHEELS

Making a wind wheel

You will need:

A large piece of thick cardboard
A plate
A pencil
A ruler
Scissors

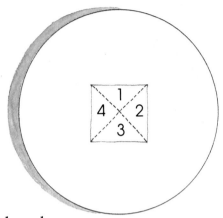

1. On the cardboard, draw a large circle, about 10 inches (25 cm) across. Cut out the circle. Use the plate as a guide for your circle.

2. Draw a square in the middle of the circle. The sides of the square should be about 2 inches (5 cm) long.

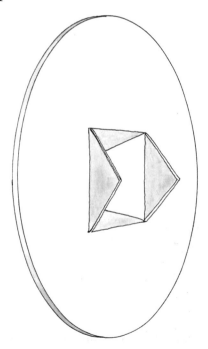

3. Draw two dotted lines diagonally on the square. Cut along these lines to make four triangles.

4. Fold triangles 1 and 3, shown above, onto one side of the wind wheel. Fold the other two triangles onto the other side.

5. Take the wheel outside and hold it upright on the ground. When the wind blows, take your hand away and watch your wind wheel move.

Further work

Make some wind wheels with triangular flaps of different sizes. With your friends, race the wheels against each other to see which ones work best.

Make a wind wheel with only one flap folded out on each side. Does it work?

Have you ever blown the seeds from a dandelion? The wind will blow the seeds farther than you can. The seeds are lighter than your wind wheel, so they will float through the air a very long way.

WIND FORCE

The **force** of the wind can be very strong. It can push over trees, it can overturn cars, and it can even blow down houses. It is often useful for us to **measure** how strong the wind is.

You can look at flags to see how strong the wind is. These flags are blowing in a strong wind.

Measuring wind force

You will need:

Wooden sticks
Strips of wrapping paper
Strips of writing paper and cloth
Pieces of string
Thumbtacks

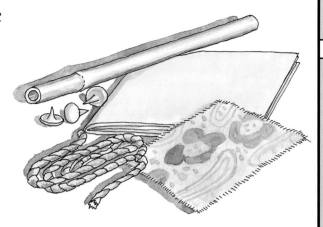

1. Use a thumbtack to attach a strip of wrapping paper to the top of a stick. When the wind is blowing, the strip will blow straight out.

2. Pin strips of writing paper, cloth, and string to other sticks. Do they blow straight out like the wrapping paper?

3. Experiment with different strips of paper, cloth, and string. Find out which is the lightest and which is the heaviest.

Making a machine for measuring wind force

You will need:

Two sticks of balsa wood, about 12 inches (30 cm) long
A junior hacksaw
Glue
*A **dowel***
A plastic bottle
Four plastic cups
Strong pins

WARNING:

Ask an adult to help you cut balsa wood.

1. With a hacksaw, cut a small piece from the center of each stick of wood. Make sure you do not cut through the wood.

2. Use some glue to fit the pieces of wood together where you have made the cuts.

3. Push a pin through the **joint** and into the end of the dowel. Use some glue to attach it more firmly.

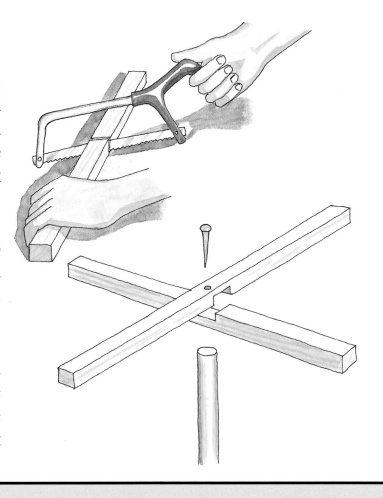

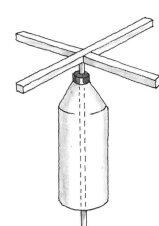

4. Make a hole in the bottom of the plastic bottle. Slide the dowel through the neck of the bottle, so that the end of the dowel sticks out of the bottom.

5. Pin a plastic cup to the end of each wooden arm. Make sure all of the cups are facing in the same direction.

6. Hold your machine outside in the wind. Count how many times the cups go around in ten seconds. Attach a colored marker to one of the cups to help you count.

7. Take your machine out on another day and count again. Is the wind stronger or weaker than before?

Machines like these measure how strong and which way the wind is blowing.

Air is what keeps a **kite** flying. It can even keep a heavy airplane up in the sky. Have you ever flown a kite on a windy day?

Do you have a kite? This one is like the kite you can make, only bigger and stronger. It should fly high on a windy day!

Making a minikite for a light breeze

You will need:

Two plastic straws   Wrapping paper
Scissors   Glue
Thread   String

1. Cut off an inch (2.5 cm) of one straw. Tie the two straws in a cross shape, as shown, and glue them together.

2. With scissors, make slots at the end of each straw. Wrap thread around the straws.

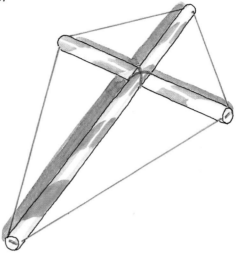

3. Cut out a piece of paper in the shape of your kite. Glue it onto the base of your kite.

4. Tie another piece of thread to the top and bottom of the longer straw. Tie a piece of string to this thread.

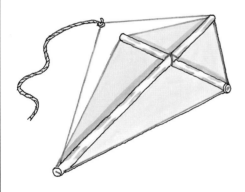

5. To make a tail, tie pieces of wrapping paper to a long piece of thread. Attach the thread to the bottom of the kite.

6. Run with the kite behind you, until you have let out all the string. This kite will only work in a very gentle breeze.

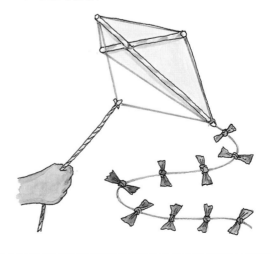

HOT-AIR BALLOONS

When air gets hot, it rises. Hot-air balloons use this warm air to float. The balloon is filled with hot air, which lifts it and makes it fly.

There are many hot-air balloons in this picture. Would you like to fly in one?

Making a model hot-air balloon

You will need:

A balloon
Thin paper
Glue
Scissors
Paints
A matchbox
Thread
Tape

1. Blow up the balloon and tie it.

2. Glue small pieces of paper onto the balloon. Cover all but the mouth of the balloon. Stick on a few layers of paper and leave the balloon to dry.

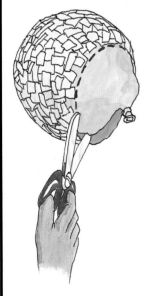

3. When the balloon is dry, cut off the bottom.

4. Paint and decorate your model. Use some thread and tape to attach the box to the bottom of your balloon. This is your passenger seat.

5. Attach a piece of thread to the top of the balloon so that you can hang it up.

PARACHUTES

Parachutes allow things to float gently through the air. People can even jump from airplanes if they have a parachute. The parachute lets them float safely to the ground.

This is a special kind of parachute. This man steers it so that it lands right where he wants it to.

Making parachutes

You will need:

Wrapping paper
Scissors
Thread
Tape
Modeling clay

1. Cut out a square of paper. Each side should be 12 inches (30 cm) long.

2. Cut out a small square hole from the center, as shown.

3. Tape the thread to each of the corners.   Make sure all four pieces of thread are exactly the same length.

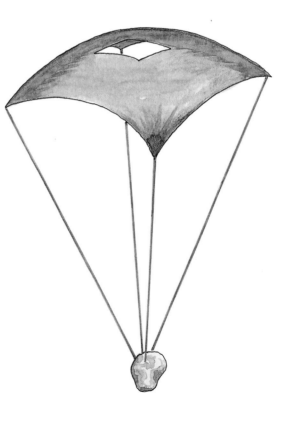

4. Attach the other ends of the thread to the modeling clay.

5. Drop your parachute from different heights. Measure the time it takes for the parachute to reach the ground each time.

6. Now make more parachutes. This time, cut several square holes out of the paper. How many squares are you able to cut out before the parachute stops working?

# JET POWER

Have you ever felt a **jet** of air? Blow up a balloon but do not tie the end. Hold the balloon and let the air out. Can you feel the air rushing out?

You have probably blown up a balloon just like this one. What happens when you let it go?

Making a balloon-powered racer

You will need:

Two strips of balsa wood, one 6 inches (15 cm) long and one 4 inches (10 cm) long
Cardboard
A balloon
A plastic straw

Tape
Glue
Strong pins
Scissors

1. Use a pin and glue to attach the two pieces of wood into a cross shape.

2. Draw three circles onto the cardboard. Cut out these shapes and use them as your wheels.

3. Attach one wheel to each end of the short piece of wood. Pin the third wheel to one side of the other piece.

4. Stretch the balloon over the straw. Tape the balloon to the straw. Use more tape to attach the balloon and straw to the racer.

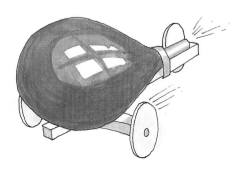

5. Blow up the balloon. Put your finger over the end to stop the air from coming out. Now put your racer on the ground and let the air out.

LAND YACHTS

A land **yacht** is like a boat with a sail, but it has wheels and travels on land instead of water.

Making a land yacht

You will need:

Five spools
A piece of balsa wood
  4 inches (10 cm)
  wide and 12 inches
  (30 cm) long
Wooden sticks
Plastic tubing
Rubber bands
Paper
Thread
Tape
Scissors

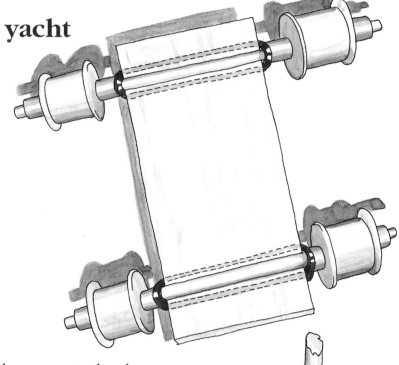

1. Use rubber bands to attach the two wooden sticks across the piece of wood.

2. On each end of the sticks, attach a spool with the tubing on either side of it. This is the base of your yacht.

3. Use two rubber bands to attach another spool upright onto the center of the base, as shown.

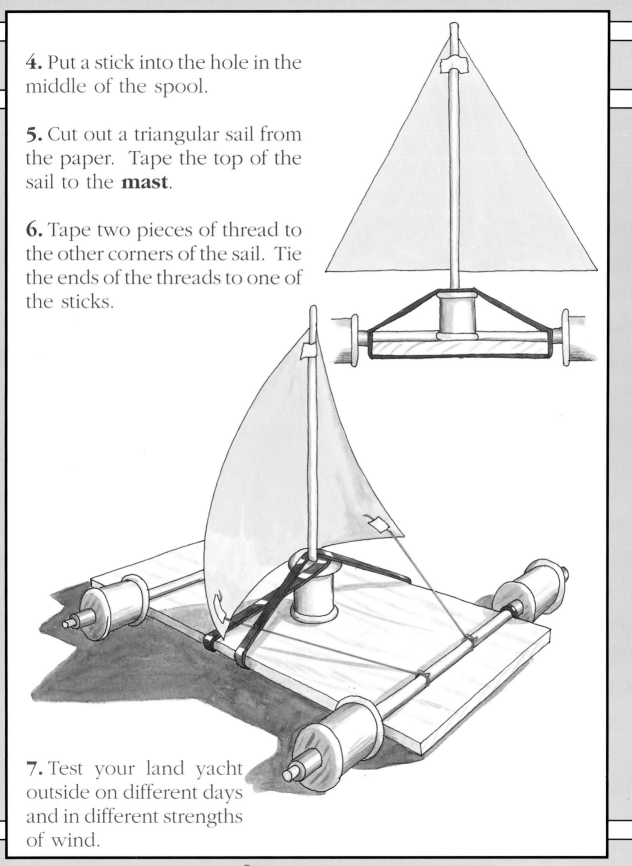

4. Put a stick into the hole in the middle of the spool.

5. Cut out a triangular sail from the paper. Tape the top of the sail to the **mast**.

6. Tape two pieces of thread to the other corners of the sail. Tie the ends of the threads to one of the sticks.

7. Test your land yacht outside on different days and in different strengths of wind.

HOVERCRAFT

Hovercraft look like big boats — but they are not boats at all! In fact, they travel on a cushion of air just above the water. Hovercraft can also be used on dry land.

This big Hovercraft can carry many people and cars.

Making a simple Hovercraft

You will need:

A Styrofoam tile
A small cardboard tube
Four cardboard strips 3/4 inch (2 cm) wide
Glue
A hair dryer

1. Make a small hole in the center of the Styrofoam tile. The hole should be the same size as the cardboard tube.

2. Set the tube into the hole and glue it in place.

3. Blow down the tube to see if your Hovercraft will rise.

4. Attach the cardboard strips to the sides of the tile, as shown. Blow down the tube now. Can you make the Hovercraft move? Ask an adult to help you use a hair drier to blow air down the tube.

> **WARNING:**
> **Do not use a hair dryer on your own. Always ask an adult to help you.**

What You'll Need

More Books About Air

Air, Air Everywhere. Tom Johnston (Gareth Stevens)
Amazing Air. Henry Smith (Lothrop)
Balloons. The Smithsonian Institution (Smithsonian)
Catch the Wind! All about Kites. Gail Gibbons (Little, Brown)
Hear the Wind Blow. Patricia Pendergraft (Scholastic)
What Makes the Wind? Laurence Santrey (Troll)
Wind and Water Energy. Sherry Payne (Raintree)
Wind Power. Norman F. Smith (Putnam)

More Books With Projects

Balloons: Building and Experimenting with Inflatable Toys. Bernie Zubrowski
  (Morrow Junior Books)
Experiments with Air. Ray Broekel (Childrens Press)
Make It with Odds and Ends! Felicia Law (Gareth Stevens)
The Scientific Kid: Projects, Experiments, Adventures. Mary S. Carson
  (Harper & Row)
Simple Science Experiments with Ping-Pong Balls. Eiji Orii and Masako Orii
  (Gareth Stevens)
Simple Science Experiments with Straws. Eiji Orii and Masako Orii (Gareth Stevens)

Places to Write for Science Supply Catalogs

White Wings
AG Industries, Inc.
3832 148th Avenue, NE
Redmond, Washington 98052

The Nature of Things
275 West Wisconsin Avenue
Milwaukee, Wisconsin 53203

Adventures in Science
Educational Insights
19560 Rancho Way
Dominguez Hills, California 90220

Suitcase Science
Small World Toys
P. O. Box 5291
Beverly Hills, California 90209

Edmund Scientific
101 East Gloucester Pike
Barrington, New Jersey 08007

Weather and Wind Instrument and
  Equipment Company
734 East Hyde Park Boulevard
Inglewood, California 90302

GLOSSARY

diagonal
Slanting from one corner of a four-sided figure, such as a square, to another corner.

dowel
A round wooden pin.

electricity
Power that travels along wires. It is used for giving light and for making all sorts of machines work.

force
The strength or power of something.

gale
A very strong wind.

jet
A thin stream of liquid or gas forced out of a small opening.

joint
The place where two or more parts come together.

kite
A toy made from a light frame covered with paper or cloth. It is flown in the wind at the end of a long string.

mast
A tall pole that holds up the sail on a boat.

measure
To find out how big or heavy something is or how quickly it moves.

parachute
A folding device shaped like an umbrella that is used to slow the fall of persons or objects from the sky.

sails
A piece of strong fabric, like canvas, that is stretched out to catch the wind.

yacht
A small, graceful ship, used for pleasure trips or racing.

Picture acknowledgements
The publishers would like to thank the following for allowing their photographs to be reproduced in this book: Eye Ubiquitous, pp. 6, 18; Chris Fairclough Colour Library, p. 11; Hutchison Library, p. 4; Topham, p. 26; Zefa, pp. 9, 12, 15, 16, 20, 22. Cover photography by Zul Mukhida.

INDEX